Lucy Muthoni

Application of Multi-Yield Curves Modelling to Kenyan Bonds Market

AF141892

Lucy Muthoni

Application of Multi-Yield Curves Modelling to Kenyan Bonds Market

LAP LAMBERT Academic Publishing

Impressum / Imprint

Bibliografische Information der Deutschen Nationalbibliothek: Die Deutsche Nationalbibliothek verzeichnet diese Publikation in der Deutschen Nationalbibliografie; detaillierte bibliografische Daten sind im Internet über http://dnb.d-nb.de abrufbar.
Alle in diesem Buch genannten Marken und Produktnamen unterliegen warenzeichen-, marken- oder patentrechtlichem Schutz bzw. sind Warenzeichen oder eingetragene Warenzeichen der jeweiligen Inhaber. Die Wiedergabe von Marken, Produktnamen, Gebrauchsnamen, Handelsnamen, Warenbezeichnungen u.s.w. in diesem Werk berechtigt auch ohne besondere Kennzeichnung nicht zu der Annahme, dass solche Namen im Sinne der Warenzeichen- und Markenschutzgesetzgebung als frei zu betrachten wären und daher von jedermann benutzt werden dürften.

Bibliographic information published by the Deutsche Nationalbibliothek: The Deutsche Nationalbibliothek lists this publication in the Deutsche Nationalbibliografie; detailed bibliographic data are available in the Internet at http://dnb.d-nb.de.
Any brand names and product names mentioned in this book are subject to trademark, brand or patent protection and are trademarks or registered trademarks of their respective holders. The use of brand names, product names, common names, trade names, product descriptions etc. even without a particular marking in this works is in no way to be construed to mean that such names may be regarded as unrestricted in respect of trademark and brand protection legislation and could thus be used by anyone.

Coverbild / Cover image: www.ingimage.com

Verlag / Publisher:
LAP LAMBERT Academic Publishing
ist ein Imprint der / is a trademark of
OmniScriptum GmbH & Co. KG
Heinrich-Böcking-Str. 6-8, 66121 Saarbrücken, Deutschland / Germany
Email: info@lap-publishing.com

Herstellung: siehe letzte Seite /
Printed at: see last page
ISBN: 978-3-659-59526-4

Zugl. / Approved by: Nairobi, University of Nairobi, 2013

TABLE OF CONTENTS

Dedication

To Cosmas. For all that you are, and what you aspire to be.

Abstract

With the onset of the financial crisis in late 2007, the market conditions changed radically and the elementary pricing procedure used among practitioners became unreliable and obsolete. This necessitated a change in the methods used in pricing financial products.

We discuss post-credit crunch paradigm shift from single-curve to the multi-curve setting[1]. We first present the single curve and explore techniques such as the construction of single yield (spot) curve, and show how we obtain yield measures from the curve and use them to price interest rate swaps. We then study the multi-curve pricing framework. This is done by constructing the single discounting curve, then the multiple forwarding curves while are used to compute forward rates and the cash flows. The discounting curve is then used to compute the discount factors and the cash flows relevant for pricing the interest rate swaps.

For our analysis, we use both single yield curve- and multi-curve pricing frameworks discussed above to price the bonds offered in Kenyan market. We use data from Central Bank of Kenya. We then compare the prices using the two methodologies of pricing, with the pricing methodology applied by the market and give conclusions and recommendations

Keywords: single-curve, Multi-curve, interest rate swaps pricing, bond pricing

---

[1] A lot of credit goes to Laursen & Bruhs (2011) whose work was exceptional and this paper would not have become a reality without it.

1. INTRODUCTION

Since the crisis of 2007-2009, one of the main focuses among practitioners is to try to come up with- and using pricing frameworks that correctly reflect the markets' practice. By undertaking this study, we hope to come up with a pricing framework that will not only be consistent with the interest-rates' market's practice, but also one that is applicable to Kenyan bond market as well.

Differences between similar rates (basis spreads) had been present in the market before the crisis, but they were generally regarded as negligible. The liquidity crisis, which reached its peak in October 2008 with the collapse of the Lehmann Brothers' Investment bank, widened the basis spreads quoted on the market between single-currency interest rate instruments, swaps in particular, characterized by different underlying rate tenors, Morini (2009).

The widened basis spreads, i.e. the difference between market rates with different underlying maturities, observed on the interest markets following the initiation of the financial crisis in 2007 indicated that the traditional pricing framework had to be revisited. To be able to correctly price interest rate swaps it became necessary to incorporate the credit and liquidity risks of different tenors, i.e. maturities, according to Ametrano & Bianchetti (2009), and consequently the pricing of swaps has become much more complex in recent years.

It is no longer adequate to use only a single yield curve when determining forward rates of different tenors, instead there is now a need for multiple curves. Furthermore, there is also a need for a different discounting practice since the pre-crisis discounting curve is no longer the best risk-free proxy. This is because the pre-crisis approach was not consistent with the market configuration during and after the crisis period.

First, it did not take into account the market information carried by the basis swap spreads, which became much larger during the period, and could no longer be ignored. Second, it did not take into account that the interest rate market had become segmented into sub-areas corresponding to different underlying rate tenors, characterized by different dynamics, Morini (2009). This meant that pricing and hedging an interest rate derivative on a single yield curve mixing different underlying rate tenors could lead to inaccurate results and more difficult to interpret. Third, by no arbitrage, discounting had to be unique; two identical future cash flows of whatever origin had to display the same present value, hence a need for a unique discounting curve.

The importance of a reliable pricing system cannot be overstated. This is because companies and financial institution frequently use interest rate derivatives for the purpose of managing interest rate risk exposure and the market has grown to contain several instruments and maturities. The most extensively used instrument that accounts for the majority of the interest rates derivatives market, and which is the focus of this study, is plain vanilla interest swaps.

As a country, Kenya is experiencing growth in the mortgage-financing sector due to the expansion of middle-class proportion of the population. As the middle-class grows, the need for housing grows too. This leads to more and more people seeking to finance their mortgages. Mortgages are offered in both fixed rates and floating rates. The latter was introduced when the market players realized that it was not realistic to lock in the interest rates due to the volatile economic atmosphere, which increased inflation rates' uncertainty. Unfortunately, the interest rate swaps are only traded over the counter which means that not all market participants have access to these tools. The other financial product whose price depends on the interest rates is the bond, which fortunately most Kenyans have access to and can use it as an investment tool.

According to Central Bank of Kenya (CBK) (2009) annual report, Kenyan Government Domestic Debt was majorly financed through the sale of Treasury bonds to commercial banks and the non-banks (61.60%), and through utilization of the overdraft facility at the Central Bank of Kenya. This means that the government could either be losing much needed funds to finance its activities if the bonds are underpriced, or could be causing financial harm to its citizens by unknowingly offering them over-priced bonds by use of wrong interest rate models (or wrong pricing process). Should we be able to come up with models that can correctly price interest rates products in Kenya, bonds included, then we will save the economy funds which can be channeled to other important developmental avenues, one of them being facilitating housing for its citizens.

The first objective of this study is to describe how interest rate swap (IRS) are priced under the single-curve pricing framework and apply this pricing framework to the Kenyan bonds' market. The second objective is to describe and apply multi-curve approach to IRS and apply the same in pricing the bonds in Kenya. Finally, we seek to compare the three approaches: single-yield curve framework, multi-curve framework and the method currently applied to price bonds in Kenya (auction method where the price is arrived at by taking the average of bid and ask prices), so that we can infer which one is the best method. The best method should be the one that correctly reflects our economic performance.

2. LITERATURE REVIEW

The pricing of interest rate swaps was, before the financial crisis in 2007, a clear cut case with a framework that researchers agreed on. In non-credit related financial literature, authors such as Ron (2000), Boenkost & Schmidt (2004) and Hull (2009) focus on bootstrapping the yield curve and determining a single forward curve such that at initiation, the present value of both legs in a swap contract equal each other. Interpolation techniques used to create smooth and continuous curves are covered by Andersen (2007) and Hagan & West (2008).

The impact of changing the discounting curve when pricing derivatives is discussed by Henrard (2007). Later, Henrard (2010) proposes a coherent valuation framework for derivatives based on different Libor tenors still using the traditional bootstrapping technique, though assuming the discounting curve as given by Ametrano & Bianchetti (2009) who assume a segmentation of the market, and bootstrap the swap rates within each tenor separately which makes their model subject to arbitrage.

An extended version of this model is suggested by Bianchetti (2010) who uses the foreign exchange analogy to prevent arbitrage opportunities. Similar approaches were followed by Chibane & Sheldon (2009) and Kijima et al (2009). Extending the theory, Mercurio (2009) builds consistent interest rate curves by modeling the joint evolution of FRA rates and implied forward rates with an extended lognormal Libor Market Model. However, this paper lacks the discussion in a multi-currency situation.

Most recently, the work of Fujii et al (2009a) and (2009b); Linderstrøm & Scavenius (2010); and Linderstrøm and Rasmussen (2011) provide a new consistent framework to construct a term structure in the presence of basis spreads and collateral in a multi-currency environment. Instead of building a yield curve by bootstrapping different liquid market instruments, now forward rates are backed out incorporating the effect of basis spreads. Johannes & Sundaresan (2007) and Whittal (2010b) extend the multi-curve pricing framework by taking the effect of collateralization on swap rates into consideration.

Kenya uses the auction model to price the bonds where the bond price is arrived at by getting the average of the bid and ask prices, Ngugi & Agoti (2009). Our contribution towards this field is the application of swaps pricing framework to Kenyan bonds market by pricing the bonds.

3. METHODOLOGY: PRICING FRAME-WORKS

3.1. DATA

We get data from Central Bank of Kenya. This data covers a period of four years, and entails both bond prices and coupon rates.

Since the data provided gives values of the coupon rates at the knot points, we use Cubic Lagrange interpolation to estimate the rates at the missing time points. This leads to attainment of the yield rates on which we apply bootstrapping to calculate the spot rates.

3.2. SINGLE-CURVE PRICING FRAMEWORK FOR IRS

The main aspect in pricing interest rate swaps prior to the financial crisis was to determine one forward/spot/discount curve. The procedure for constructing the swap curve was generally agreed on among the practitioners, though there existed no single correct methodology. Furthermore, practitioners faced the challenge in regard to which interpolation technique to apply, as well as incorporating turn of the year effect into the chosen curve. Essentially, different approaches with respect to these challenges resulted in varying forward and discounting curve.

Bianchetti (2009) Summarize the traditional single-curve pricing and hedging framework in the following steps:

3.2.1. Select one finite set of vanilla (i.e. basic) interest rate instruments with increasing maturities.

3.2.2. Build one spot curve using the selected instruments and bootstrapping (a method for building the curve incrementally in increasing maturity order)

3.2.3. Compute on the same curve, forward rates and discount factors and work out the prices by summing up the discounted cash flows.

3.2.4. Compute the delta sensitivity, i.e. how the prices react to changes in interest rates, and hedge the resulting delta risk using the suggested amounts (hedge ratios) of the same set of vanillas.

Recall that in the single-curve framework, a single curve is used for both discounting and forwarding. By single-curve, it is meant that the same instruments are used to derive all three curves; the discount curve, the spot curve and the forward curve.

All the curves can be derived from each other and there is no need to specify exactly which curve is referred to when the term single-curve is used. In literature, however, single-curve often refers to the spot curve, and so we adapt this view.

We have now defined the three fundamental curves used in swap pricing and the relationship between them.

3.2.1. The theoretical framework

In this section, the theoretical framework is examined for the single-curve framework. There exist several official benchmarks for interbank term deposits such as Libor, Euribor, Cibor or Tibor. The spot Libor rate is defined as the rate of return from buying 1 unit of a default free zero-coupon bond at time t and selling it at maturity T_n. Hence, the spot Libor rate is in fact the discounting rate:

$$L_{t,T_n} = \frac{1}{\delta_n}\left(\frac{1}{P_{t,T_n}} - 1\right) \qquad (3.1.)$$

Here, P_{t,T_n} refers to the default free discounting factor where δ_n is the daycount fraction for the interval $[t, T_n]$. Then, the forward Libor rate from T_{n-1} to T_n standing at time t can be estimated by the following equation:

$$F_{t;T_{n-1},T_n} = \frac{1}{\delta_n} \left(\frac{P_{t,T_{n-1}}}{P_{t,T_n}} - 1 \right) \tag{3.2}$$

Here, δ_n is now the daycount fraction for the interval $[T_{n-1,T_n}]$. Due to the relation between forward Libor rates and discounting factors in equation 3.2 the single-curve framework avoids arbitrage opportunities.

An instrument that is based on the forward Libor rate is the Forward Rate Agreement (FRA). Taking a long position in a FRA, the payoff at maturity T_n can be determined by the difference between the spot Libor rate and the fixed rate K:

$$V_{T_n} = \delta_n \left(L_{T_{n-1},T_n} - K \right) \tag{3.3}$$

Determining the value of the FRA at time t the following equation is applied:

$$V_t = \delta_n \left(E_t \left[L_{T_{n-1},T_n} \right] - K \right) \tag{3.4}$$

This imposes the challenge to determine the forward Libor rate. Here, $E_t [\ ]$ denotes the expectations operator under the T_n-forward measure Q^{T_n}. At initiation, the FRA rate K is determined such that it sets both legs to par:

$$\delta_n \, K P_{t,T_n} = \delta_n \left(E_t \left[L_{T_{n-1},T_n} \right] P_{t,T_n} \right) \tag{3.5}$$

The choice of a zero-coupon bond maturing at time T_n as numeraire is particularly useful when dealing with interest rate derivatives. It follows that any simply compounded forward rate spanning a time interval, ending in T_n, is a martingale under the T_n-forward measure, i.e.:

$$F_{t;T_{n-1},T_n} = E_t \left[L_{T_{n-1},T_n} \right] \tag{3.6}$$

By applying this relation it follows that equation 3.2 can be written as:

$$F_{t;T_{n-1},T_n} = E_t\left[L_{T_{n-1},T_n}\right] = \frac{1}{\delta_n}\left(\frac{P_{t,T_{n-1}}}{P_{t,T_n}} - 1\right) \tag{3.7}$$

3.2.2. Pricing IRS using different rates

Derivatives dependent on future interest rates can be priced by applying forward rates, which becomes a straightforward procedure, given equation 3.7 above. Simultaneously, an interest rate swap (IRS) can be priced as a portfolio of several FRAs where both legs of the swap must also equal each other at initiation:

$$\text{IRS: } C_n \sum_{n=1}^{N} \Delta_n P_{t,T_n} = \sum_{n=1}^{N} \delta_n \left(E_t\left[L_{T_{n-1},T_n}\right]P_{t,T_n}\right) \tag{3.8}$$

Here, C_n is the par swap rate of the N-length IRS at time t, Δ_n and δ_n are the day-count fractions of the fixed and floating leg, respectively. For simplicity it is assumed that the payments of the fixed and floating leg occur simultaneously. Inserting equation 3.7 into equation 3.8 yields:

$$C_n \sum_{n=1}^{N} \Delta_n P_{t,T_n} = \sum_{n=1}^{N} \delta_n \left(\frac{1}{\delta_n}\left(\frac{P_{t,T_{n-1}} - P_{t,T_n}}{P_{t,T_n}}\right)\right) P_{t,T_n}$$

$$C_n \sum_{n=1}^{N} \Delta_n P_{t,T_n} = \sum_{n=1}^{N}\left(P_{t,T_{n-1}} - P_{t,T_n}\right)$$

$$C_n \sum_{n=1}^{N} \Delta_n P_{t,T_n} = P_{t,T_0} - P_{t,T_n} \tag{3.9}$$

The right hand side of the above equation can be considered as a long position in one zero-coupon bond with maturity T_0 and a short position in another zero-coupon bond with maturity T_N. Finally, the swap rate can be determined as:

$$C_N = \frac{P_{t,T_0} - P_{t,T_n}}{\sum_{n=1}^{N} \Delta_n P_{t,T_n}} \qquad (3.10)$$

Naturally, the swap rate C_N must be equal to the rate on the swap curve with maturity N. When considering a tenor swap (TS) the present values of both floating legs must likewise equal each other at initiation. The TS can be seen as a portfolio of two IRS of the same maturity with matching fixed legs and two floating legs plus a tenor spread added to the floating leg that is indexed to the shorter tenor. Hence, the required relation between the two floating legs is:

$$TS: \sum_{n=1}^{N} \delta_n \left(E_t \left[L_{T_{n-1},T_n} \right] + \tau_N \right) P_{t,T_n} = \sum_{m=1}^{M} \delta_m E_t \left[L_{T_{m-1},T_m} \right] P_{t,T_m} \quad (3.11)$$

Where τ_N denotes the time t market spread of the length N between the two underlying Libor rates with tenors n < m. The left hand side could resemble the 3M underlying Libor rate whereas the right hand side could reflect the 6M underlying Libor rate.

In this example, the 3M Libor payer compensates the higher credit risk inherent in the 6M Libor rate by adding the 3M vs. 6M tenor basis spread. Solving for τ_N such that both legs equal each other at initiation yields:

$$\tau_N = \frac{\sum_{m=1}^{M} \delta_m E_t \left[L_{T_{m-1},T_m} \right] P_{t,T_m} - \sum_{n=1}^{N} \delta_n \left(E_t \left[L_{T_{n-1},T_n} \right] P_{t,T_n} \right)}{\sum_{n=1}^{N} \delta_n P_{t,T_n}} \qquad (3.12)$$

Bianchetti (2009) consider the above as being the difference in the two swap rates from two plain interest rate swaps with different underlying tenors but the same maturity $T_N = T_M$:

$$\tau_N = C_M = CT_N \qquad (3.13)$$

The two approaches expressed in equation 3.12 and 3.13 to determine the tenor basis spread might differ slightly from each other. This is due to different day-count conventions, as the fixed leg usually is paid on 30/360 basis whereas the spread is added to the floating leg that is based on (act days)/360. Here, dealers have different reasons for trading contracts applying the two different approaches, Linderstrom (2011).

In the case of a cross currency swap (CCS) the interest rate payments of both legs occur in different currencies. From the possible types of CSSs: fixed vs. fixed, fixed vs. floating and floating vs. floating, the latter type is particular important and is used for generating the other types synthetically.

Assuming that both legs have the same tenor but depend on different underlying rates a CCS must satisfy the following relation:

$$\text{CCS:} \left(-P_{t,T_0}^f + \sum_{n=1}^N \delta_n^f \left(E_t^f\left[L_{T_{n-1},T_n}^f\right] + b_N\right) P_{t,T_n}^f + P_{t,T_N}^f\right) fx_t =$$
$$-P_{t,T_0} + \sum_{n=1}^N \delta_n \; E_t \left[L_{T_{n-1},T_n}\right] P_{t,T_n} + P_{t,T_N} \tag{3.14}$$

Here, the index f denotes that the variable is relevant for a foreign currency where b_N is the basis spread for length N such that the US Dollar as base currency trades flat against the foreign currency. E_t^f [ ] denotes the expectations operator under the T_n-forward measure $Q_f^{T_n}$ in the foreign currency applying P_{t,T_N}^f as Numeraire.

The spot exchange rate of US Dollar per foreign currency at time t is represented by fx_t. The US Dollar acts as a base currency but could easily be replaced by another base currency. Similarly, to equation 3.12, determining the cross currency basis spread can be done by isolating b_N in equation 3.4 which yields:

$$b_N = \frac{\left(P^f_{t,T_0} + \sum_{n=1}^{N} \delta^f_n \left(E^f_t\left[L^f_{T_{n-1},T_n}\right]\right) P^f_{t,T_n} - P^f_{t,T_N}\right) fx_t}{\sum_{n=1}^{N} \delta_n P_{t,T_n}}$$

$$+ \frac{\left(-P_{t,T_0} + P_{t,T_N} + \sum_{n=1}^{N} \delta_n \left(E_t\left[L_{T_{n-1},T_n}\right]\right)\right) P_{t,T_n}}{\sum_{n=1}^{N} \delta_n P_{t,T_n}}$$

(3.15)

The no-arbitrage condition from equation 3.2 creates the foundation for pricing IRS, TS and CCS as in equations 3.8, 3.11 and 3.14, respectively. Consequently, determining the forward curve enables practitioners to estimate swap rates, tenor basis spreads and cross-currency basis spreads.

3.2.3. Choice of Numeraire

The following section takes its form from the work of German (1995), Mercurio (2006), White (2009) and Bjork (2009). In general, a Numeraire is a reference asset that is chosen in a way to normalize all other asset prices with respect to it. The bank account is often used as a risk neutral Numeraire. However, this is just one of many possible choices since any positive, non-dividend paying asset can be applied as Numeraire. Dealing with interest rate derivatives, the choice of a zero-coupon bond as Numeraire is particularly useful.

3.2.3.1.   Martingales

Defining a sequence of random variables X_0, X_1, . . . ,X_t, the variable X_t is a martingale if, for all $t > 0$, the following is true:

$$E[X_t|X_{t-1}, X_{t-2}, \ldots, X_0] = X_{t-1}$$

(3.16)

Similarly, a variable θ is a martingale if it follows a zero-drift stochastic process:

$$d\theta = \sigma\, d\mathcal{Z} \tag{3.17}$$

Where $d\mathcal{Z}$ is a Wiener process. The volatility parameter σ can be considered a stochastic variable itself or it can depend on θ as well as on other stochastic variables. The convenience of the martingale property is shown in its tremendous applicability in financial literature, where the expected value at any future time is equal to its value today:

$$E[\theta_T] = \theta_t \tag{3.18}$$

The change in θ between time t and time T is the sum of the changes over many small time intervals. Consequently, the expected change must be zero.

The equivalent martingale measure result

The equivalent martingale measure result shows that under the no-arbitrage condition the relationship between two price processes is a martingale. White (2009) introduces the market price of risk λ as:

$$\lambda = \frac{\mu - r}{\sigma} \tag{3.19}$$

Here, μ and σ are the return and volatility on θ, respectively, and r is the risk free rate. The market price of risk measures the risk adjusted excess return with respect to securities that depend on θ.

Furthermore, the market price of risk must at any given time be the same for all derivatives that are dependent only on θ and t to ensure no arbitrage. From White (2009) equation 3.19 only holds for investment assets that provide no income.

Assuming the two prices of traded assets X and N depend on a single source of uncertainty and provide no income during the time of matter, the relationship between the prices of the two assets is a martingale for some choice of the market price of risk. The relationship is defined by:

$$\phi = \frac{X}{N} \tag{3.20}$$

This can be thought of as measuring the price of X in units of N, where the security price of N is referred to the Numeraire. Choosing the same market price of risk for instruments X as for a given Numeraire N makes the relative price ϕ a martingale. To prove this result, the price processes of X and N are defined as:

$$dX = (r + \lambda\sigma_X)Xdt + \sigma_X XdZ \tag{3.21}$$

$$dN = (r + \lambda\sigma_N)Ndt + \sigma_N NdZ \tag{3.22}$$

Where μ is replaced by rewriting equation 3.19. Choosing the same market price of risk for X as for N, i.e. $\lambda = \sigma_X = \sigma_N$, results in a zero-drift relative price process ϕ as written below:

$$d\phi = (\sigma_X - \sigma_N)\phi \, dZ \tag{3.23}$$

This is similar to equation 3.6 where now the process of ϕ is a martingale.

3.2.4. Deriving the Swap Curve

3.2.4.1.   Zero-coupon bond as Numeraire

Assuming there exists a Numeraire N and a probability measure Q^N, equivalent to the initial risk neutral measure Q^0, the price of any traded asset X relative to N is a martingale under Q^N, i.e.:

$$\frac{X_t}{N_t} = E^N \left[\frac{X_T}{N_T} | \mathcal{F}t \right] \qquad (3.24)$$

This proposition introduced by (German, 1995) provides a fundamental tool for pricing derivatives as it is generally applicable for any positive non-dividend paying Numeraire.

When applying the zero-coupon bond as Numeraire any simple compounded forward rate is a martingale under the T-forward measure. Hence, the price of an interest rate derivative π_t at time t can be estimated as the discounted expected payoff on claim H at maturity T, conditional on the T-forward measure:

$$\pi_t = P_{t,T} E^T \left[H_T | \mathcal{F}t \right] \qquad (3.25)$$

The reason why the measure Q^T is called forward measure is justified in the following. Recalling equation 3.2 and rewriting it gives:

$$F_{t;Tn-1,Tn} P_{t,Tn} = \frac{1}{\delta_n} \left(P_{t,Tn-1} - P_{t,Tn} \right) \qquad (3.26)$$

Now, considering $F_{t;Tn-1,Tn} P_{t,Tn}$ as a traded asset X_t and by applying $P_{t,Tn}$ as the numeraire N_t the left hand side (LHS) of equation 3.24 yields:

$$\text{LHS:} \frac{F_{t;Tn-1,Tn} P_{t,Tn}}{P_{t,Tn}} = \frac{1}{\delta_n} \left(\frac{P_{t,Tn-1}}{P_{t,Tn}} - 1 \right) = F_{t;Tn-1,Tn} \qquad (3.27)$$

Similarly, the right hand side (RHS) of equation 3.24 can be rewritten as:

$$\text{RHS:} E_t \left[\frac{F_{t;Tn-1,Tn} P_{TnTn,}}{P_{t,Tn}} | \mathcal{F}_t \right] = E_t \left[F_{Tn-1,Tn-1,Tn} \right] \qquad (3.28)$$

In the following the filtration $\mathcal{F}_t$ is omitted for simplification purposes. Finally, replacing $F_{Tn-1,Tn-1,Tn}$ with its equivalent L_{T_{n-1},T_n} on the right hand side of the above equation and setting equations 3.27 and 3.28 equal to each other yields:

$$F_{t;T_n-1,T_n} = E_t\left[L_{Tn-1,Tn}\right] \tag{3.29}$$

This corresponds exactly to equation 3.6 applied to price interest rate swaps and hence indirectly FRAs.

3.3. MULTI-CURVES PRICING FRAMEWORK

3.3.1. Introduction

As we have seen in the previous section, the interest rate market condition changes dramatically in the second half of 2007 and several anomalies could be observed leading to a necessity to use different pricing framework. As in the single-curve framework, the curves in the multiple-curve framework have three different forms: the discount curve (curve of zero-coupon bond prices), the spot curve and the forward curve (the instantaneous forward rate curve).

However, instead of having one curve of each form, the multiple-curve approach has different curves depending on the underlying tenor of the bootstrapping instrument. The most common tenors are 1M, 3M, 6M and 12M, which means that it is common to construct four different spot and/or forward curves. These curves, however, are not used for discounting. Instead, a discount curve is constructed separately with different bootstrapping instruments than those in the construction of the forward curves.

According to Ametrano & Bianchetti (2009), the multiple-curve framework for pricing and hedging interest rate derivatives (hence, interest rate swaps) can be summarized as follows:

3.3.2. Construct a single discounting curve

3.3.3. Select different sets of vanilla interest rate instruments, each with homogeneous underlying rate tenors (e.g. 1M, 3M, 6M, 12M)

3.3.4. Construct multiple separated forwarding curves, one for each tenor, using the selected instrument and bootstrapping.

3.3.5. Compute forward rates on each forwarding curve as well as the cash flows relevant for pricing derivatives on the same underlying.

3.3.6. Compute discount factor using the discounting curve and sum up the discounted cash flows in order to work out prices

3.3.7. Compute the delta sensitivity and hedge the delta risk with appropriate hedge ratio of the corresponding set of vanilla instruments.

3.3.2. FRA and Implied Forward Rates

The forward rate $F_{t;T_{n-1},T_n}$ can be estimated from two spot rates $R_{t,T_{n-1}}$ and R_{t,T_n} in the following way:

$$F_{t;T_{n-1},T_n} = [\frac{(1+R_{t,T_n})^{T_n}}{(1+R_{t,T_{n-1}})^{T_{n-1}}}]^{\frac{1}{\delta_n}} - 1 \qquad (3.30)$$

Here, δ is the day-count fraction for the interval $[T_{n-1},T_n]$.

Designated banks that contribute to the Libor panels are selected to be the upper part of the banking world in terms of credit standing, reputation and activity in the cash markets. A population that was considered virtually risk less before the crisis Morini

(2009, p. 5). Thus, the Libor rate L_{t,T_n} was regarded as a good approximation to the risk free spot rate R_{t,T_n}. The equation for the forward rate is then:

$$F_{t;T_{n-1},T_n} = [\frac{(1+L_{t,T_n})^{T_n}}{(1+L_{t,T_{n-1}})^{T_{n-1}}}]^{\frac{1}{\delta_n}} - 1 = \frac{1}{\delta_n}(\frac{P_{t,T_{n-1}}}{P_{t,T_n}} - 1)$$ (3.31)

Furthermore, the payoff of a long position in a FRA contract at maturity T_n is known to be:

$$V_{T_n} = \delta_n(L_{T_{n-1},T_n} - K).$$ (3.32)

The FRA is quoted through its equilibrium rate:

$$F_{t;T_{n-1},T_n} = E_t[L_{T_{n-1},T_n}]$$ (3.33)

This in theory should correspond to the FRA rate K, making such a deal fair at initiation.

3.3.3. Constructing a Single Discounting Curve/Swap Curve

3.3.3.1. Swap curve construction without collateral

The large discrepancies between the market FRA rates and their implied forward rates during the financial crisis have a significant impact on the theoretical framework for pricing interest rate derivatives when applying the standard replication method. Formerly, it was commonly agreed on that placing money over a 12M period should yield the same as investing in two 6M deposits. However, this is no longer the case, making the following equation no longer viable.

$$F_{t;T_{n-1},T_n} = \frac{1}{\delta_n}\left(\frac{P_{t,T_{n-1}}}{P_{t,T_n}} - 1\right)$$ (3.34)

Following the notation introduced in the pre-crisis framework, m and n correspond to semiannual and quarterly payments for the underlying Libor rates. Likewise, the swap rate, the tenor spread of a 3M vs. 6M tenor swap and the cross-currency basis spread where the US Dollar as base currency trades flat against the foreign currency f, all with maturity N, are denoted as C_N, τ_N and b_N, respectively. Consequently, the required conditions for a bank located in the US that identifies the 3M US Dollar Libor rate as its appropriate discounting rate are given as follows:

$$\text{IRS: } C_M \sum_{m=1}^{M} \Delta_m P_{t,T_m} = \sum_{m=1}^{M} \delta_m E_t \left[L_{T_{m-1},T_m} \right] P_{t,T_m} \tag{3.35}$$

$$\text{TS: } \sum_{n=1}^{N} \delta_n \left(E_t \left[L_{T_{n-1},T_n} \right] + \tau_n \right) P_{t,T_n} = \sum_{m=1}^{M} \delta_m E_t \left[L_{T_{m-1},T_m} \right] P_{t,T_m} \tag{3.36}$$

$$\text{CCS} = -P_{t,T_0} + \sum_{n=1}^{N} \delta_n E_t \left[L_{T_{n-1},T_n} \right] P_{t,T_n} + P_{t,T_n}$$

$$= \left(-P_{t,T_0} \sum_{n=1}^{N} \delta_n^f \left(E_t^f \left[L_{T_{n-1},T_n}^f \right] + b_N \right) P_{t,T_N}^f \right) f x_t \tag{3.37}$$

Here, it is assumed that N = 2M and fx_t symbolizes the exchange rate of US Dollar per foreign currency at time t. We rewrite equation (3.33) as:

$$P_{t,T_0} - P_{t,T_n} + \tau_n \sum_{n=1}^{N} \delta_n, P_{t,T_n} = \sum_{m=1}^{M} \delta_m E_t \left[L_{T_{m-1},T_m} \right] P_{t,T_m} \tag{3.38}$$

Eliminating the floating legs from the above equation and equation 3.31 yields:

$$C_M \sum_{m=1}^{M} \Delta_m, P_{t,T_m} - \tau_n \sum_{n=1}^{N} \delta_n, P_{t,T_n} = P_{t,T_0} - P_{t,T_n} \tag{3.39}$$

Once the set of $\{P_t, T\}$ is sequentially estimated and properly interpolated, the set of 3M US Dollar forward Libor rates $\{E_t[L_{T_{n-1,n}}]P_{t,T_n}\}$ can be recovered from equation 3.34. Similarly, the set of 6M US Dollar forward Libor rates $\{E_t \left[L_{T_{m-1,T_m}} \right] P_{t,T_m}\}$

can be backed out from equation 3.35. It is possible to extend this procedure by adding more tenor swaps to derive additional forward curves for tenors such as 1M and 12M. As the 3M US Dollar Libor rate is assumed to be the discounting rate, the LHS of the equation has to be zero and thus can be rewritten to:

$$P_{t,T_0}^f - P_{t,T_N}^f - b_N \sum_{n=1}^N \delta_n^f P_{t,T_n}^f = \sum_{n=1}^N \delta_n^f E_t^f \left[L^f_{T_{n-1,T_n}} \right] P_{t,T_N}^f \qquad (3.40)$$

Moreover, the constraint from a EUR IRS is:

$$C_N^f \sum_{n=1}^N \delta_n^f P_{t,T_n}^f = \sum_{n=1}^N \delta_n^f \sum_{n=1}^N \delta_n^f E_t^f \left[L^f_{T_{n-1,T_n}} \right] P_{t,T_N}^f \qquad (3.41)$$

By eliminating the floating parts from equations 3.39 and 3.40, the following equation can be used to extract the set of foreign discounting factors $\{P_{t,T_N}^f\}$ sequentially as all other inputs are quoted in the market:

$$\left(C_N^f + b_N \right) \sum_{n=1}^N \delta_n^f P_{t,T_n}^f = P_{t,T_0}^f - P_{t,T_N}^f \qquad (3.42)$$

Applying appropriate spline methods gives a continuous discounting curve. Now, the set of 3M Euro forward Libor rates $\{ E_t^f \left[L^f_{T_{n-1,T_n}} \right] \}$ can be obtained from equation 3.40 by using the determined discounting factors. It is possible to extend this procedure by adding more Euro tenor swaps to derive additional forward curves for tenors such as 1M, 6M and 12M. Having a domestic Libor as discounting rate, the cross-currency basis spread does not affect the US Dollar discounting factors as can be seen from equation 3.38, whereas the Euro discounting factors depend not only on the EUR IRS quotes but also on the basis spreads in the EUR/USD CCS.

For foreign banks which have their funding bases in the US market and thus still apply a US Dollar rate as discounting rate, the same procedure will be carried out under the assumption that the 3M US Dollar Libor is the discounting rate from the perspective of a foreign bank. Now, the initial conditions for IRS, TS, and CCS can be reformulated as:

$$\text{IRS: } C_M^f \sum_{m=1}^{M} \Delta_M^f \, P_{t,T_m}^f = \sum_{m=1}^{M} \delta_M^f \, E_t^f \left[L_{T_{m-1},T_m}^f \right] P_{t,T_m}^f \tag{3.43}$$

$$\text{TS: } \sum_{n=1}^{N} \delta_n^f \left(E_t^f \left[L_{T_{n-1},T_n}^f \right] + \tau_N^f \right) P_{t,T_N}^f = \sum_{m=1}^{M} \delta_M^f \, E_t^f \left[L_{T_{m-1},T_m}^f \right] P_{t,T_m}^f \tag{3.44}$$

$$\text{CCS: } = -P_{t,T_0} + \sum_{n=1}^{N} \delta_n \, E_t \left[L_{T_{n-1},T_n} \right] P_{t,T_n} + P_{t,T_n}$$

$$= \left(-P_{t,T_0}^f \sum_{n=1}^{N} \delta_n^f \left(E_t^f \left[L_{T_{n-1},T_n}^f \right] + b_N \right) P_{t,T_N}^f + P_{t,T_N}^f \right) fx_t \tag{3.45}$$

It can be seen that the RHS of equations 3.43 and 3.44 are equivalent, hence:

$$C_M^f \sum_{m=1}^{M} \Delta_M^f \, P_{t,T_m}^f = \sum_{n=1}^{N} \delta_n^f \left(E_t^f \left[L_{T_{n-1},T_n}^f \right] + \tau_n^f \right) + P_{t,T_N}^f \right)$$

$$C_M^f \sum_{m=1}^{M} \Delta_M^f \, P_{t,T_m}^f - \sum_{n=1}^{N} \delta_n^f \tau_n^f \, P_{t,T_N}^f = \sum_{n=1}^{N} \delta_n^f \left(E_t^f \left[L_{T_{n-1},T_n}^f \right] P_{t,T_N}^f \right) \tag{3.46}$$

As the 3M US Dollar Libor rate is still assumed to be the discounting rate, the LHS of equation 3.45 has to be zero and thus can be rewritten as:

$$\sum_{n=1}^{N} \delta_n^f \left(E_t^f \left[L_{T_{n-1},T_n}^f \right] + b_N \right) P_{t,T_N}^f = P_{t,T_0}^f - P_{t,T_N}^f$$

$$\sum_{n=1}^{N} \delta_n^f E_t^f \left[L_{T_{n-1},T_n}^f \right] P_{t,T_N}^f + \sum_{n=1}^{N} \delta_n^f b_N P_{t,T_N}^f = P_{t,T_0}^f - P_{t,T_N}^f \tag{3.47}$$

Inserting equation 3.46 into 3.47 and further simplifying yields:

$$C_M^f \sum_{m=1}^{M} \Delta_M^f \, P_{t,T_m}^f - \sum_{n=1}^{N} \delta_n^f (b_N - \tau_n^f) P_{t,T_N}^f = P_{t,T_0}^f - P_{t,T_N}^f \tag{3.48}$$

To derive the continuous set of Euro discounting factors $\{P^f_{t,T_N}\}$ from this formula by appropriate splining is possible. Consecutively, the set of 6- and 3M Euro forward Libor rates, i.e. $\{E^f_t[L^f_{T_{m-1,T_m}}]\}$ and $\{E^f_t[L^f_{T_{n-1,T_n}}]\}$, can be calculated from equations 3.35 and 3.36, respectively. Following this approach under the assumption of the 3M US Dollar Libor rate being the discounting rate, the Euro discounting and two Euro forward curves can be extracted, enabling the foreign bank to consistently mark-to-market EUR IRS with EUR TS and EUR/USD CCS at the same time. Further TS conditions could be added to this framework in order to extract forward curves for other tenors. Fujii et al (2009b) highlight the relation among discounting and tenor-dependent yield curves. Following their work, the term Δfm Pft, Tm can be approximated by:

$$\Delta^f_M \, P^f_{t,T_m} \simeq \frac{\Delta^f_M}{2}\left(P^f_{t,T_m-3m} + P^f_{t,T_m}\right) \tag{3.49}$$

Now, simplifying equation 3.48 by replacing $\Delta^f_M = 2\Delta^f_n$ yields

$$\sum_{n=1}^N \Delta^f_n C^f_M + \delta^f_n\left(b_N - \tau^f_n\right)P^f_{t,T_n} \simeq P^f_{t,T_0} + P^f_{t,T_N} \tag{3.50}$$

This approximation indicates that the effective swap rate, meaning the swap rate that correctly accounts for tenor and cross-currency basis spreads, differs from the normal swap rate in the following relation:

$$C^f_{M\,eff} \simeq C^f_M + \frac{\delta^f_n}{\Delta^f_n}\left(b_N - \tau^f_n\right) \tag{3.51}$$

Consequently, having a position in a plain vanilla interest rate swap exposes the holder inevitably to tenor and cross-currency basis spreads.

3.3.3.2. Swap curve construction with collateral

Swap markets experienced tremendous growth in the late 1980's and early 1990's, when an increasingly diverse group of counterparties entered the markets. Consequently, market practitioners developed a number of credit enhancements to improve the quality of the swap contracts, seeking to mitigate the exposure to counterparty risk. Johannes and Sundaresan (2007) agree on the most important credit enhancement being the posting of collateral in the amount of the current mark to- market value of the swap contract. Overall, 70 percent of all OTC derivatives transactions were subject to collateral agreements in the year 2010 compared to only 30 percent in 2003 (ISDA, 2010, p.10).

As noted in Piterbarg (2010), collateralized contracts are based on the credit support annex (CSA) to the International Swaps and Derivatives Association master agreement. That is why these contracts are often referred to as CSA trades. In a collateralized swap contract, the collateral is posted from the counterparty to the institution that has a positive present value of the contract. To compensate the counterparty putting up collateral, the institution needs to pay the margin called collateral rate on the outstanding collateral to the payer. The legal collateral agreement between the two parties can vary substantially in the independent amount, minimum transfer amount and threshold amount. The independent amount must be posted independently of the exposure between the two parties whereas the threshold amount is the lower bound before additional collateral is required.

The minimum transfer amount is required to reduce operational burdens. Additionally, Linderstrøm and Rasmussen (2011) comment on the difficulties pricing such contracts when asymmetries in the above described amounts occur and refer to credit valuation adjustments (CVA) when pricing such contracts. As collateral is used to offset liabilities in case of a default, it can be considered a risk free investment why the collateral rate can be seen as a proxy for the risk free rate Piterbarg (2010, p.97).

Pricing collateralized contracts faces the challenge of asymmetry that arises from the credit risk. In order to deal with this problem, Fujii et al (2009b) assume perfect and continuous collateralization in cash with a threshold amount of zero, i.e. collateral is posted continuously and the posted amount of cash is 100 percent of the contract's present value. This simplification allows neglecting the counterparty default risk and recovers the symmetry in the collateral payments. Moreover, this allows the decomposition of the cash flow of a collateralized swap into a portfolio of independently collateralized strips of payments.

Considering a stochastic process of the collateral account V (t) with an appropriate self-financing strategy under the risk neutral measure, one could invest the posted collateral at the risk free interest rate but would need to pay the collateral rate. The process of the collateral account is then given by:

$$dV (s) = y(s)V(s)ds + a(s)dh(s) \qquad (3.52)$$

where, $y(s) = r(s) - c(s)$ is the difference of the risk free rate $r(s)$ and the collateral rate $c(s)$ in the domestic currency at time s, $h(s)$ is the time s value of the derivative maturing at T with the cash flow $h(T)$ and $a(s)$ is the number of positions of the derivative. Integrating equation 3.52 yields the following:

$$V (T) = e^{\int_t^T y(u)du} V (t) + \int_t^T e^{\int_s^T y(u)du} a(s)dh(s) \qquad (3.53)$$

The trading strategy of perfect and continuous collateralization with a threshold of zero is given such that:

$$V(t) = h(t)$$

$$a(s) = e^{\int_t^s y(u)du} \tag{3.54}$$

Applying the above trading strategy to equation 3.50 determines the value of the collateral account at time T:

$$V(T) = e^{\int_t^T y(s)ds} h(T) \tag{3.55}$$

The present value of the underlying derivative h(t) is then given by the following:

$$h(t) = E_t^Q\left[e^{-\int_t^T r(s)-y(s)ds} h(T)\right] = E_t^Q\left[e^{-\int_t^T c(s)ds} h(T)\right] \tag{3.56}$$

Here, E_t^Q [ ] denotes the expectation operator under the money-market account Q at time t. From this result it is clear that the discounting of future cash flows of collateralized trades is done at the collateral rate and not the Libor rate. Especially in distressed markets, the Libor rate for a corresponding currency can differ significantly from the collateral rate. Therefore, it follows that Libor discounting is not appropriate for the pricing of collateralized trades.

For some trades, collateral can alternatively be posted in a foreign currency. The procedure for determining at what factor discounting should be done is very similar to the procedure described previously, though subject to modifications. Considering the process of the collateral account V^f

$$dV^f(s) = y^f(s)V^f(s)ds + a(s)d\left[\frac{h(s)}{fx(s)}\right] \tag{3.57}$$

The foreign exchange rate at time s is given by $f_x(s)$ where $y_f(s) = r_f(s) - c_f(s)$ is the difference of the risk free rate $r_f(s)$ and the collateral rate $c_f(s)$ of the foreign currency f at time s. Again, integrating the process for the collateral account gives:

$$V^f(T) = e^{\int_t^T y^f(s)ds} V^f(t) + \int_t^T e^{\int_s^T y^f(u)du} a(s)d \begin{bmatrix} h(s) \\ \overline{fx(s)} \end{bmatrix} \tag{3.58}$$

Now, the trading strategy for collateral posted in a foreign currency can be adopted:

$$V^f(T) = \begin{bmatrix} h(t) \\ \overline{fx(t)} \end{bmatrix}$$

$$a(s) = e^{\int_t^s y^f(u)du} \tag{3.59}$$

This reduces equation 3.58 to:

$$V^f(T) = e^{\int_t^T y^f(s)ds} \frac{h(t)}{fx(t)} \tag{3.60}$$

Then, the present value of the underlying derivative in terms of the domestic currency is given by:

$$h(t) = V^f(t)fx(t) = E_t^Q \left[e^{-\int_t^T r(s)ds} V^f(t)fx(t) \right]$$

$$= E_t^Q \left[e^{-\int_t^T r(s)ds} e^{\int_t^T (r^f(s) - c^f(s))} h(T) \right] \tag{3.61}$$

Again, from the above equation it becomes clear that for contracts where collateral is posted in a foreign currency, Libor discounting is inappropriate. The difference from the previous example is now that the collateral earns the foreign risk free interest rate less the foreign collateral rate as opposed to the domestic rates.

In order to price collateralized swaps it is critical to determine the discounting curve based on an overnight rate. Here, it is very convenient to use the OIS rate already quoted in the market. By assuming that the OIS rate is continuously and perfectly collateralized with a zero threshold and approximating the daily compounding with continuous compounding, the condition from the OIS can be derived from equation 3.56 such that:

$$S_N \sum_{n=1}^{N} \Delta_n E_t^Q \left[e^{-\int_t^{T_n} c(s)ds} \left(e^{-\int_{T_{n-1}}^{T_n} c(s)ds} - 1 \right) \right] \tag{3.62}$$

Here, S_N denotes the time t par rate for the length N OIS, where $c(t)$ is the overnight rate, i.e. the collateral rate, at time t. Now, by defining the discounting factor as:

$$D_{t,T} = E_t^Q \left[e^{-\int_t^T c(s)ds} \right] \tag{3.63}$$

Equation 3.59 reduces to:

$$\text{OIS: } S_N \sum_{n=1}^{N} \Delta_n D_{t,T_n} = D_{t,T_0} - D_{t,T_n} \tag{3.64}$$

Obtaining the continuous set of discounting factors $\{D_{t,T}\}$ can be done by appropriate splining. Deriving the discounting factors from the collateral rate assumes that the OIS market is available up until the necessary maturity.

3.3.4. Deriving Discounting Curves

Uncollateralized derivatives should be discounted by each financial institution, using its own unique funding cost. This entails discussions whenever two counterparties have a different perception of the other's funding cost and therefore the following part will focus solely on collateralized swaps neutralizing the ambiguity of discounting.

If the US Dollar collateral rate is equal to the risk free rate then discounting is done at the risk free rate for any given currency. The set of discounting factors $\{D_{t,T}\}$ and $\{D_{t,T}^f\}$ (where the former denotes US Dollar and the latter the foreign discounting factor, i.e. for Euro) are estimated using OIS discounting rates as opposed to the riskier LIBOR rates. For this purpose, we have:

$$D_{t,T_N} = \frac{1-S_N \sum_{n=1}^{N-1} \delta_n D_{t,T_N}}{1+S_N \delta_N} \tag{3.65}$$

$$D_{t,T_N}^f = \frac{1-S_N^f \sum_{n=1}^{N-1} \delta_n^f D_{t,T_n}^f}{1+S_N^f \delta_n^f} \tag{3.66}$$

3.4. KENYAN BOND MARKETS

3.4.1. Introduction to Kenyan Bond Market

The Kenyan bonds market traces its origin back to the 1980s when the Government of Kenya first launched a bid to use treasury bonds as a source of funds to finance government deficit. The first corporate bond was issued on 8 November 1996 by the East African Development Bank (EADB), which issued a multi-lateral bond. This market experienced a turn-around in 2001 however, when the government re-launched treasury bonds. In both cases, the rates applicable were floating rates pegged to the 91-day Treasury bill rates Ngugi and Agoti (2009).

By end of 2005, there were 65 Treasury bonds (Floating Rate, Special and Fixed Rate Bonds) and 5 corporate bonds listed on the Nairobi Securities Exchange. There has been a tremendous market gain in terms of the size of the bond market. Total bond value has increased from just over Kshs. 0.8 billion in 1996, to over 186 billion in 2003, with the coupon rate in the market ranging between 8% and 14% Ngugi and Agoti (2009).

3.4.2. Trading system and Pricing of the Bonds

Bonds trading at the NSE follow a call auction system to determine the bond price. In a call auction, the orders are accumulated periodically and are matched at a specific time at the price, which results in maximum trading volume. In most cases, this price is the average price between ask and bid price. Additionally, given that Nairobi Securities Exchange (NSE)'s bond market does not have market makers, it is order driven and this according to Madhavan (2001) reduces efficiency as compared to if it were a quote driven market Ngugi and Agoti (2009). It is clear that the pricing should be done in such a way that it takes into consideration the global market practices.

4. RESULTS AND DISCUSSION

4.1. SINGLE - CURVE PPRICING

4.1.1. Numerical Results and Discussion

We get the results indicated in appendix A1: SINGLE YIELD CURVE PRICING FRAME-
WORK NUMERICAL RESULTS

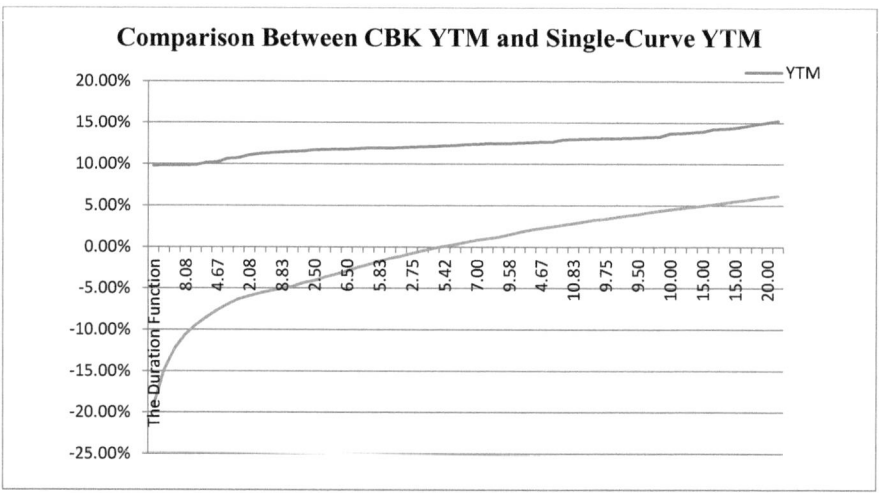

We see that using the single-curve framework to calculate the forward rates, results
in negative values at some points. Though it is rare to get such rates in real life, it
comes as no surprise given the method used to price bonds in Kenya. Bonds trading
at the NSE follow a call auction system to determine the bond price. In a call
auction, the orders are accumulated periodically and are matched at a specific time
at the price, which results in maximum trading volume. In most cases, this price is
the average price between ask and bid price see 3.2.2: Trading system and Pricing of the

Bonds. This would explain why there would be such a mismatch between the rates we have generated and those ones applicable in reality.

Also, as we move down the table, we notice that the rates tend towards 6%, which is a more stable and logical figure, compared to rates above 12% which in reality would fuel demand-driven inflation. This shows that there is need for the Central Bank of Kenya in conjunction with Nairobi Securities Exchange to engage Financial Mathematicians so that they can assist in coming up with better and reflective bond prices.

4.2. MULTI-CURVE PRICING FRAMEWORK

4.2.1. Methodology

We are going to follow Ametrano & Biancheti (2009) outline, which can be summarized as at section 3.3 MULTI-CURVES PRICING FRAMEWORK:

Step 1: In coming up with the discount curve: we have interpolated the 2011 CBK rates given in the A3: CENTRAL BANK OF KENYA: BONDS DATA FOR THE YEAR 2011.

Step 2: Obtaining Spot Curve: This is done by getting discount factors bootstrapped from the market quotes, given in

A2: BOND DATA FROM CBK ANNUAL REPORT 2009 then they are bootstrapped to get spot rates and forward rates from with which we can calculate the discounted cash flows.

Step 3: In calculating the Discount Factors, we use of the following equation which was suggested by Ametrano & Biancheti (2009). The time factor was assumed to be one:

$$P(t,T) = (1/(1+\text{Discount Rate}*\text{Time factor})) \tag{4.1}$$

Step 4: Then we calculate the forward rates and the spot rates, using the following equations:

$$\mathbf{F_S(t: T_1, T_2)} = -\frac{\ln P(t,T_2) - \ln P(t,T_1)}{\delta(T_1,T_2)} \tag{3.2}$$

$$R(t,T) = -\frac{\ln P(t,T)}{\delta(t,T)} \tag{4.2}$$

Which assist us in calculating the yield rates, using the following equation Fabozzi (2004):

$$\text{Par Value} = \text{Present Value} (1 + y)^n \tag{4.3}$$

Where y are the yield rates

4.2.2. Numerical Results and discussion

The numerical results are shown in A4: MULTI-YIELD CURVE PRICING FRAME-WORK NUMERICAL RESULTS.

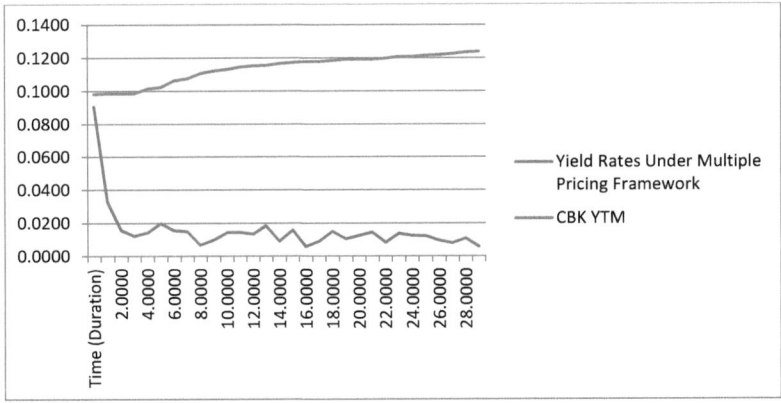

We see that using the multi-curve pricing framework, we are able to get much lower interest rates to be used to price the bonds. The curve is very different from the one currently used, and this is because of the difference in the pricing methodologies.

4.2.3. Comparison between the three practices: Single-Curve, Multi-Curve and Auction (Currently used in Kenya)

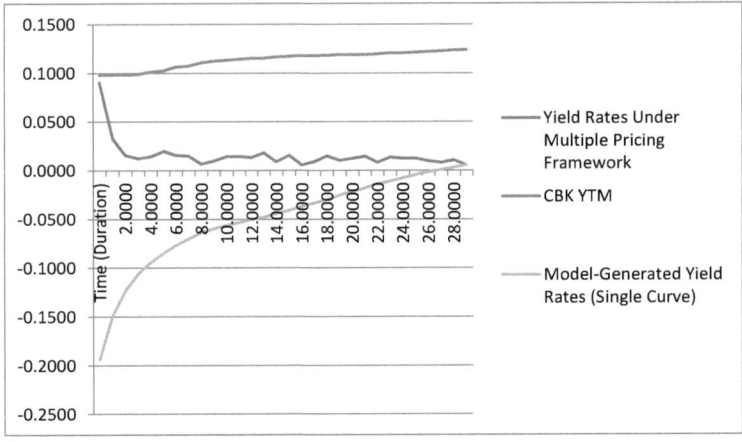

We see that with the single-curve model, we are getting negative yield rates at some points, which is not realistic. With the multiple curves, we get interest rates which are more reflective of the current market's interest behavior. This is because the country's interest rates have not been going up continuously, as the two other pricing frameworks indicate.

5. CONCLUSION AND RECOMMENDATION

In our study, we set out to describe how interest rate swap (IRS), under the single-curve pricing framework, are priced. We also sought to compare this pricing framework with the one currently applied in the Kenyan bonds' market. We have looked into both single-curves pricing framework to price the swaps theoretically, and applied the framework to price Kenyan bonds. Using this framework, we were able to get a curve with positive gradient all through. However, we see that this method produced negative interest rates, which is not realistic. When compared to the framework currently used to price Kenyan bonds, we see that the single-curves pricing framework might not be better than the one being currently used because it gives us negative interest rates.

The second objective was to describe and apply multi-curve approach to IRS and apply the same in pricing the bonds (since we do not have access to interest rate swaps market in Kenya) by calculating the yield rates. Using multi-curve pricing framework, we were able to find that the results were more reflective of the interest rates movement. We see a graph that depicts a volatile economic environment by the upwards and downward movement of the interest rates.

The third objective was to compare the three approaches: single-yield curve framework, multi-curve framework and the method currently applied to price them, to infer which one is more realistic. Looking at the three graphs each representing a different pricing framework, we can conclude that the improved approach is more consistent with the market situation in that it offers a more reflective picture of how interest rates have been trending, though it showed much lower rates compared to the ones in the market. This could be explained by the fact that the pricing practice used today has no technical explanation, since it depends on the market's forces of demand and supply. Even though multi-curve pricing framework seems to be better than the other two (single-curve, which gives negative values; and auction, which shows an unrealistic trends in interest rate movements), it demands much more additional efforts.

First, the discounting curve clearly plays a special and fundamental role and must be built with particular care. While the forwarding curve construction is driven by the underlying rate tenor homogeneity principle, for which there is a general market consensus, there is no longer general consensus for the discounting curve construction. Some practitioners use forward rates curve while others use the discount curve. When it comes to the financial product to be used, some use OIS while others use different curves. Second, building multiple curves requires multiple quotations: much more interest rate bootstrapping instruments must be considered (deposits, futures, swaps, basis swaps, FRAs, etc.), which are available on the market with different degrees of liquidity and can display transitory inconsistencies.

In particular, looking at the Kenyan market, we have very few products which can be used for the multiple-curve pricing framework. As we undertook this study, we realized the only products that can give us different underlying products with different tenors would be fixed deposit rates with different maturities. In Kenya, there is an urgent need for introduction of interest rate swaps which would be used to address the concerns mentioned in the significance of this study, given that we have seen that our current pricing framework gives an incorrect picture of interest rates trends.

REFERENCES

1. Agoti, R. N. (2009). Microstructure Characteristics of the Bond Market in Kenya. African Finance Journal.

2. Andersen, L. (2007). "Discount curve construction with tension splines,". In Review of Derivatives Research (Vol. 10(3), pp. 227–267).

3. Bianchetti, F. A. (2009). Bootstrapping the Illiquidity: Multiple Curves Construction for Market Coherent Forward Rate Estimation. Financial Journal.

4. Bianchetti, M. (2010). "Two Curves, One Price: Pricing & Hedging Interest Rate Derivatives Decoupling Forwarding and Discounting Yield Curves,". In Quantitative Finance Papers.

5. Bjork, T. (2009). Arbitrage Theory in Continuous Time. Stockholm: Oxford, Finance.

6. Boenkost, W., & Schmidt, W. (2004). Cross currency swap valuation. HfB.

7. Bruhs, M. S. (2011). Pricing of Interest Rate Swaps in the Aftermath of the Financial Crisis.

8. Central Bank of Kenya. (2009). Annual Report.

9. Chibane, M., & Sheldon, G. (2009). "Building curves on a good basis,". SSRN eLibrary.

10. Fabozzi., F. (2004). Mathematics of Financial Modeling and Investment Management. Wiley and Sons.

11. Filipovic, D. (2009). Term Structure Models, A Graduate Course. Vienna, Austria: Springer.

12. Fujii, M., Shimada, Y., & Takahashi, A. (2009a). "A Market Model of Interest Rates with Dynamic Basis Spreads in the presence of Collateral and Multiple currencies.

13. Fujii, M., Shimada, Y., & Takahashi, A. (2009b). "A note on construction of multiple swap curves with and without collateral,".

14. German, E. K. (1995). Changes of numeraire, changes of probability measure and option pricing. Journal of Applied Probability, 32(2), 443-458.

15.Hagan, P., & West, G. (2008, May). "Methods for constructing a yield curve,". WILMOTT Magazine, 70–81.

16.Henrard, M. (2007). "The irony in the derivatives discounting,". Wilmott Magazine, 92, 98.

17.Henrard, M. (2010). "The Irony in Derivatives Discounting Part II: The Crisis,". Wilmott journal, 2(6), 301–316.

18.Hull, J. (2009). Options, futures and other derivatives. Pearson Prentice Hall.

19.Johannes, M., & Sundaresan, S. (2007). "The impact of collateralization on swap rates,". In The Journal of Finance (Vol. 62(1), pp. 383–410).

20.Kijima, M., Tanaka, K., & T.Wong. (2009). "A multi-quality model of interest rates,". In Quantitative Finance (Vol. 9(2), pp. 133–145).

21.Laursen, B. (2011). Pricing Interest Rate Swaps in the Aftermath of the Financial Crisis. Aarhus University, Aarhus School of Business.

22.Linderstrøm, M., & Rasmussen, N. (2011, February, (1)). "Kompleksiteten i simple rentederivater. Finans Invest.

23.Linderstrøm, M., & Scavenius, O. (2010). "From curves to surfaces: How plain vanilla grew complex,". In Danske Markets at Den Danske Finansanalytikerforening.

24.Linderstrom, R. (2011). Kompleksiteten i simple rentederivater. Finans Invest.

25.Mercurio, D. a. (2006). Interest Rate Models: Theory and Practice. Springer.

26.Mercurio, F. (2009). "Interest rates and the credit crunch: new formulas and market models,". In Preprint. Bloomberg, QFR.

27.Morini, M. (2009). "Solving the puzzle in the interest rate market (Part 1 & 2),".

28.Ron, U. (2000). A practical guide to spot curve construction.

29.Whittall, C. (2010b). "Dealing with Funding on Uncollateralised Swaps,". In Risk.net (p. 25).

APPENDICES

A1: SINGLE YIELD CURVE PRICING FRAME-WORK NUMERICAL RESULTS

Time Frame	The Duration Function	YTM (GOK Rates)	Price/Par Value	Forward Rates	Spot Rates	Present Value	YTM (Calculated Rates
May 2006 - July 2011	5.17	9.77%	100.07	-0.1935	-0.1935	124.0868	-0.1935
July 2004 - Aug 2011	7.08	9.82%	99.61	0.0007	-0.1017	137.4972	-0.1489
Aug 2003 to Sept 2011	8.08	9.84%	99.35	0.0003	-0.0689	147.2849	-0.1230
Feb 2009 - Sept 2011	2.58	9.84%	99.72	-0.0014	-0.0535	156.1875	-0.1061

43

June 2005 - Dec 2011	6.50	9.85%	101.4	-0.0025	-0.0457	166.4310	-0.0943
May 2007 - Jan 2012	4.67	10.14%	100.58	0.0017	-0.0372	171.4676	-0.0851
Feb 2010 - Feb 2012	2.00	10.22%	98.75	0.0093	-0.0233	172.3702	-0.0765
Aug 2004 - Mar 2012	7.58	10.63%	97.92	0.0011	-0.0185	174.1426	-0.0694
Feb 2010 - Mar 2012	2.08	10.73%	97.47	0.0022	-0.0134	175.6953	-0.0634
Sept 2009 - May 2012	2.67	11.04%	101.36	-0.0144	-0.0253	187.4449	-0.0596
June 2006 - June 2012	6.00	11.20%	100.48	0.0015	-0.0220	189.9878	-0.0563
Sept 2003 - July 2012	8.83	11.31%	98.25	0.0026	-0.0178	189.1467	-0.0531
May 2007 - Aug 2012	5.17	11.43%	97.96	0.0006	-0.0159	191.6417	-0.0503

May 2007 - Sept 2012	5.33	11.51%	97.75	0.0004	-0.0144	194.0204	-0.0478
Feb 2010 - Sept 2012	2.50	11.53%	91.36	0.0280	0.0142	178.7926	-0.0438
June 2006 - Nov 2012	6.42	11.66%	99.78	-0.0132	0.0018	194.9224	-0.0410
Feb 2010 - Dec 2012	2.83	11.72%	90.53	0.0361	0.0387	170.2581	-0.0365
July 2006 - Jan 2013	6.50	11.77%	102.03	-0.0173	0.0214	187.8594	-0.0333
May 2008 - Jan 2013	4.67	11.77%	96.83	0.0115	0.0338	172.4503	-0.0299
Feb 2011 - Feb 2013	2.00	11.83%	90.39	0.0356	0.0713	150.2715	-0.0251
June 2007 - Apr 2013	5.83	11.90%	99.33	-0.0154	0.0553	156.4842	-0.0214
May 2008 - Apr 2013	4.92	11.90%	96.16	0.0067	0.0628	142.5328	-0.0177

Date Range							
Feb 2011 - Apr 2013	2.08	11.90%	92.89	0.0169	0.0813	127.3382	-0.0136
Oct 2010 - June 2013	2.75	11.97%	102.14	-0.0329	0.0461	133.8534	-0.0112
Oct 2010 - Aug 2013	2.83	12.05%	93.49	0.0327	0.0806	113.3789	-0.0077
May 2008 - Aug 2013	5.25	12.06%	95.26	-0.0035	0.0771	107.2526	-0.0046
May 2008 - Oct 2013	5.42	12.13%	94.81	0.0009	0.0784	98.9848	-0.0016
July 2006 - Dec 2013	7.42	12.18%	99.61	-0.0065	0.0717	97.0381	0.0009
Aug 2006 - Feb 2014	7.50	12.23%	102.19	-0.0034	0.0684	93.1800	0.0032
July 2007 - July 2014	7.00	12.34%	93.55	0.0132	0.0827	78.7832	0.0057
May 2005 - Sept 2014	9.33	12.37%	92.56	0.0011	0.0842	71.8938	0.0082

Aug 2007 - Feb 2015	7.50	12.44%	100.86	-0.0110	0.0726	73.0404	0.0101
Sept 2006 - Apr 2015	9.58	12.46%	103.02	-0.0022	0.0704	69.6955	0.0119
May 2010 - May 2015	5.00	12.47%	83.39	0.0471	0.1210	50.3238	0.0150
May 2010 - Nov 2015	5.42	12.53%	80.63	0.0063	0.1283	43.1240	0.0180
May 2011 - Jan 2016	4.67	12.58%	83.18	-0.0066	0.1211	39.6819	0.0208
Oct 2006 - Mar 2016	9.42	12.62%	104.74	-0.0219	0.0968	45.5581	0.0228
Oct 2006 - May 2016	9.58	12.66%	104.71	0.0000	0.0970	41.5185	0.0246
Nov 2006 - Sept 2017	10.83	12.93%	103.39	0.0012	0.0984	37.3215	0.0265
Oct 2007 - Oct 2017	10.00	12.94%	90.72	0.0140	0.1139	29.3986	0.0286

Oct 2008 - Feb 2018	9.33	12.99%	90.23	0.0006	0.1147	26.2306	0.0306
Oct 2008 - July 2018	9.75	13.04%	89.62	0.0007	0.1156	23.3528	0.0325
Dec 2006 - Aug 2018	11.67	13.05%	104.27	-0.0120	0.1023	24.6480	0.0341
Oct 2008 - Sept 2018	9.92	13.06%	89.36	0.0168	0.1210	18.8433	0.0360
Oct 2009 - Apr 2019	9.50	13.12%	88.59	0.0009	0.1222	16.6474	0.0378
Dec 2007 - May 2019	11.42	13.13%	99.33	-0.0095	0.1116	16.7910	0.0394
Oct 2010 - Apr 2020	9.50	13.21%	77.37	0.0299	0.1450	11.4229	0.0415
Oct 2010 - Oct 2020	10.00	13.25%	79.23	-0.0023	0.1424	10.2395	0.0435
Mar 2007 - Mar 2022	15.00	13.65%	104.64	-0.0162	0.1240	12.0316	0.0451

Jun 2007 - Jun 2022	15.00	13.71%	98.78	0.0040	0.1285	10.0642	0.0467
Nov 2007 - Nov 2022	15.00	13.81%	92.54	0.0045	0.1337	8.3165	0.0484
Mar 2008 - Mar 2023	15.00	13.89%	92.03	0.0004	0.1342	7.2919	0.0500
Oct 2009 - Oct 2024	15.00	14.18%	90.01	0.0015	0.1360	6.2780	0.0515
Mar 2010 - Mar 2025	15.00	14.25%	76.15	0.0121	0.1499	4.6190	0.0533
Dec 2010 - Dec 2025	15.00	14.34%	67.76	0.0083	0.1594	3.5449	0.0551
Jun 2008 - Jun 2028	20.00	14.60%	94.68	-0.0142	0.1430	4.3334	0.0566
May 2011 - May 2031	20.00	14.82%	69.34	0.0183	0.1640	2.7264	0.0584
May 2010 - May 2035	25.00	15.04%	75.55	-0.0033	0.1603	2.5603	0.0601

Jan 2011 - Jan 2041	30.00	15.25%	78.94	-0.0014	0.1587	2.3088	0.0617

A2: BOND DATA FROM CBK ANNUAL REPORT 2009

Issue No.	Maturity Date	Coupon	YTM	CBK Price	NSE Price
FXD2/2006/5	25/07/2011	11.25%	9.77%	**100.07**	100.2
FXD2/2004/7	15/08/2011	7.00%	9.82%	**99.61**	99.79
FXD2/2003/8	19/09/2011	7.00%	9.84%	**99.35**	99.56
FXD3/2009/2	19/09/2011	8.75%	9.84%	**99.72**	99.93
FXD1/2005/6	19/12/2011	13.00%	9.85%	**101.4**	101.38
FXD1/2007/5	23/01/2012	11.25%	10.14%	**100.58**	100.59
FXD1/2010/2	30/01/2012	8.00%	10.22%	**98.75**	98.78
FXD1/2004/8	12/3/2012	7.50%	10.63%	**97.92**	98.08
FXD2/2010/2	26/03/2012	7.11%	10.73%	**97.47**	97.67
FXD1/2003/9	14/05/2012	12.75%	11.04%	**101.36**	101.72
FXD1/2006/6	18/06/2012	11.75%	11.20%	**100.48**	100.95
FXD2/2003/9	16/07/2012	9.50%	11.31%	**98.25**	98.79
FXD2/2007/5	20/08/2012	9.50%	11.43%	**97.96**	98.61
FXD3/2007/5	17/09/2012	9.50%	11.51%	**97.75**	98.4

FXD3/2010/2	24/09/2012	3.81%	11.53%	**91.36**	91.93
FXD2/2006/6	19/11/2012	11.50%	11.66%	**99.78**	100.48
FXD4/2010/2	24/12/2012	4.59%	11.72%	**90.53**	91.04
FXD1/2006/7	21/01/2013	13.25%	11.77%	**102.03**	102.56
FXD1/2008/5	21/01/2013	9.50%	11.77%	**96.83**	97.31
FXD1/2011/2	25/02/2013	5.28%	11.83%	**90.39**	90.73
FXD1/2007/6	22/04/2013	11.50%	11.90%	**99.33**	99.4
FXD2/2008/5	22/04/2013	9.50%	11.90%	**96.16**	96.22
FXD2/2011/2	22/04/2013	7.44%	11.90%	**92.89**	92.94
FXD1/2003/10	10/6/2013	13.25%	11.97%	**102.14**	102.32
FXD2/2003/10	12/8/2013	8.50%	12.05%	**93.49**	93.75
FXD3/2008/5	19/08/2013	9.50%	12.06%	**95.26**	95.54
FXD4/2008/5	21/10/2013	9.50%	12.13%	**94.81**	95.18
FXD2/2006/7	16/12/2013	12.00%	12.18%	**99.61**	100.11
FXD1/2006/8	17/02/2014	13.25%	12.23%	**102.19**	102.82
FXD1/2007/7	21/07/2014	9.75%	12.34%	**93.55**	94.37
FXD1/2009/5	15/09/2014	9.50%	12.37%	**92.56**	93.44
FXD1/2007/8	16/02/2015	12.75%	12.44%	**100.86**	108.13
FXD1/2006/9	13/04/2015	13.50%	12.46%	**103.02**	112.75

FXD1/2010/5	18/05/2015	6.95%	12.47%	**83.39**	93.17
FXD2/2010/5	23/11/2015	6.67%	12.53%	**80.63**	80.57
FXD1/2011/5	25/01/2016	7.64%	12.58%	**83.18**	94.95
FXD1/2006/10	14/03/2016	14.00%	12.62%	**104.74**	118.86
FXD2/2006/10	16/05/2016	14.00%	12.66%	**104.71**	114.13
FXD1/2006/11	11/9/2017	13.75%	12.93%	**103.39**	125.85
FXD1/2007/10	16/10/2017	10.75%	12.94%	**90.72**	112.21
FXD1/2008/10	12/2/2018	10.75%	12.99%	**90.23**	112.98
FXD2/2008/10	16/07/2018	10.75%	13.04%	**89.62**	117.57
FXD1/2006/12	13/08/2018	14.00%	13.05%	**104.27**	133.45
FXD3/2008/10	17/09/2018	10.75%	13.06%	**89.36**	116.06
FXD1/2009/10	15/04/2019	10.75%	13.12%	**88.59**	101.28
FXD1/2007/12	13/05/2019	13.00%	13.13%	**99.33**	111.92
FXD1/2010/10	13/04/2020	8.79%	13.21%	**77.37**	77.9
FXD2/2010/10	19/10/2020	9.31%	13.25%	**79.23**	103.1
FXD1/2007/15	7/3/2022	14.50%	13.65%	**104.64**	141.58
FXD2/2007/15	6/6/2022	13.50%	13.71%	**98.78**	135.57
FXD3/2007/15	7/11/2022	12.50%	13.81%	**92.54**	120.59
FXD1/2008/15	13/03/2023	12.50%	13.89%	**92.03**	118.55

FXD1/2009/15	7/10/2024	12.50%	14.18%	**90.01**	127.73
FXD1/2010/15	10/3/2025	10.25%	14.25%	**76.15**	107.66
FXD2/2010/15	8/12/2025	9.00%	14.34%	**67.76**	85.82
FXD1/2008/20	5/6/2028	13.75%	14.60%	**94.68**	134.3
FXD1/2011/20	5/5/2031	10.00%	14.82%	**69.34**	72.07
FXD1/2010/25	28/05/2035	11.25%	15.04%	**75.55**	70.91
SDB1/11/30	21/01/2041	12.00%	15.25%	**78.94**	81.03

A3: CENTRAL BANK OF KENYA: BONDS DATA FOR THE YEAR 2011

Duration	Yield Rate
1 year	21.41
2 years	14.61
5 years	10.64
10 years	11.77
12 years	16.64
15 years	12.39
20 years	14.40
30 years	14.29

This figures have been arrived at by getting the average of the bonds, where we might have more than one bonds offered for one given duration

A4: MULTI-YIELD CURVE PRICING FRAME-WORK NUMERICAL RESULTS

AMETRANO BIANCHETTI (2009) METHOD		
Step One: Coming up with the discount curve.	Step Two: Obtaining Spot Curve	Step Three: Calculating the Discount Factors.

Discount Curve	YTM	Forward Rates	Spot Rates	Discount Factor, P(t,T)	Price/Par Value	Spot rate (using the relationship described in Equation 3.1.7 with the	Forward Rates (using the relationship described in Equation 3.1.10	New Present Values: Multi-curve set-up	Yield Rates Under Multiple Pricing Framework

	assumption that time frame =1								
0.09065	91.33	0.1940	0.1940	99.61	0.8237	‑0.1017	0.0007	0.0982	0.21
0.03303	93.10	0.0593	0.1364	99.35	0.8725	‑0.0689	0.0003	0.0984	0.15
0.01582	95.13	0.0349	0.1020	99.72	0.9030	‑0.0535	‑0.0014	0.0984	0.11
0.01222	96.59	0.0067	0.0954	101.4	0.9090	‑0.0457	‑0.0025	0.0985	0.10
0.01436	94.42	‑0.0057	0.1011	100.58	0.9038	‑0.0372	0.0017	0.1014	0.11
0.01983	90.13	‑0.0137	0.1149	98.75	0.8914	‑0.0233	0.0093	0.1022	0.12
0.01546	91.07	0.0241	0.0911	97.92	0.9129	‑0.0185	0.0011	0.1063	0.10

0.01494	90.05	-0.0015	0.0926	97.47	0.9115	- 0.0134	0.0022	0.1073	0.10
0.00677	95.43	0.0067	0.0859	101.36	0.9177	- 0.0253	-0.0144	0.1104	0.09
0.00987	91.92	-0.0251	0.1113	100.48	0.8947	- 0.0220	0.0015	0.112	0.12
0.01425	86.78	-0.0304	0.1421	98.25	0.8675	- 0.0178	0.0026	0.1131	0.15
0.01447	85.34	-0.0117	0.1539	97.96	0.8573	- 0.0159	0.0006	0.1143	0.17
0.01330	85.40	0.0043	0.1496	97.75	0.8610	- 0.0144	0.0004	0.1151	0.16
0.01828	78.68	0.0144	0.1353	91.36	0.8735	0.0142	0.0280	0.1153	0.14
0.00902	88.62	0.0186	0.1168	99.78	0.8898	0.0018	-0.0132	0.1166	0.12
0.01563	79.12	0.0203	0.0967	90.53	0.9078	0.0387	0.0361	0.1172	0.10
0.00581	91.90	0.0134	0.0834	102.03	0.9200	0.0214	-0.0173	0.1177	0.09
0.00899	86.31	0.0017	0.0817	96.83	0.9216	0.0338	0.0115	0.1177	0.09
0.01490	76.56	-0.0154	0.0972	90.39	0.9073	0.0713	0.0356	0.1183	0.10
0.01050	82.28	-0.0366	0.1345	99.33	0.8741	0.0553	-0.0154	0.119	0.14

0.15	0.119	0.0067	0.0628	0.8663	96.16	0.1436	-0.0090	78.38	0.01234
0.16	0.119	0.0169	0.0813	0.8585	92.89	0.1525	-0.0089	73.76	0.01457
0.16	0.1197	-0.0329	0.0461	0.8584	102.14	0.1527	-0.0002	83.82	0.00831
0.18	0.1205	0.0327	0.0806	0.8460	93.49	0.1672	-0.0145	73.19	0.01367
0.19	0.1206	-0.0035	0.0771	0.8423	95.26	0.1716	-0.0043	74.49	0.01241
0.19	0.1213	0.0009	0.0784	0.8412	94.81	0.1730	-0.0014	73.95	0.01221
0.19	0.1218	-0.0065	0.0717	0.8431	99.61	0.1707	0.0023	78.36	0.00959
0.18	0.1223	-0.0034	0.0684	0.8488	102.19	0.1639	0.0068	81.19	0.00797
0.16	0.1234	0.0132	0.0827	0.8591	93.55	0.1519	0.0121	74.23	0.01081
0.00	0.1237	0.0011	0.0842	1.0000	92.56	0.0000	0.1640	85.37	0.00575

Printed by Books on Demand GmbH, Norderstedt / Germany